THIS NOTEBOOK BELONGS TO: ________________

AF887851

Geometry Transformations

Transformation: means to change a given geometric shape.

Pre-Image: the original figure before a transformation.

Image: the resulting figure after the transformation.

Translation: slide the pre-image in any direction without changing its shape, size, or orientation.

Example: Translate the given triangle to the right 5 units.

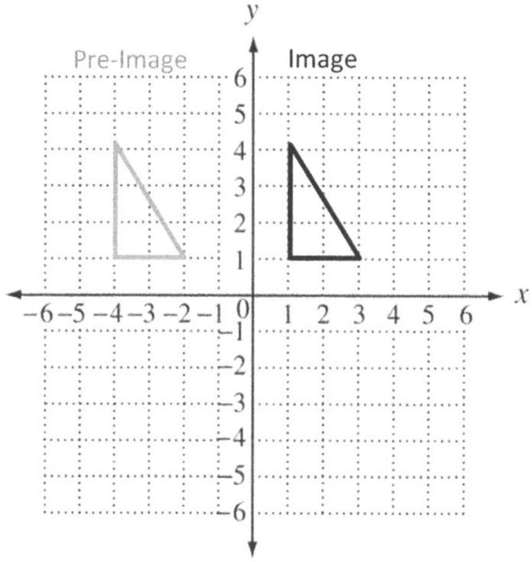

Reflection: flip the pre-image over a central line, also called a mirror line.

Example: Reflect the given triangle about the *x*-axis.

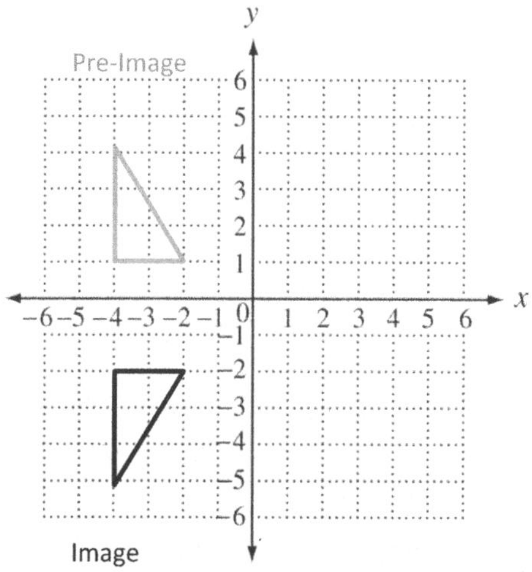

Rotation: turn or rotate the pre-image by a certain degree around a center point.

Example: Rotate the given triangle 90° about the vertex $(-4,1)$.

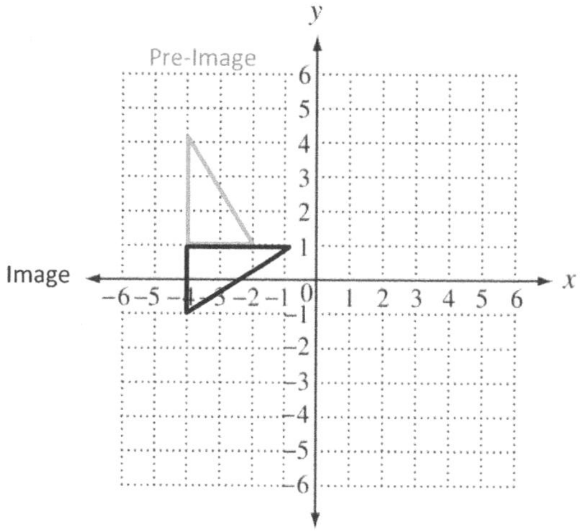

Glide Reflection: the image is formed by a reflection and translation.

Example: Reflect the given triangle about the *y*-axis and translate down 3 units.

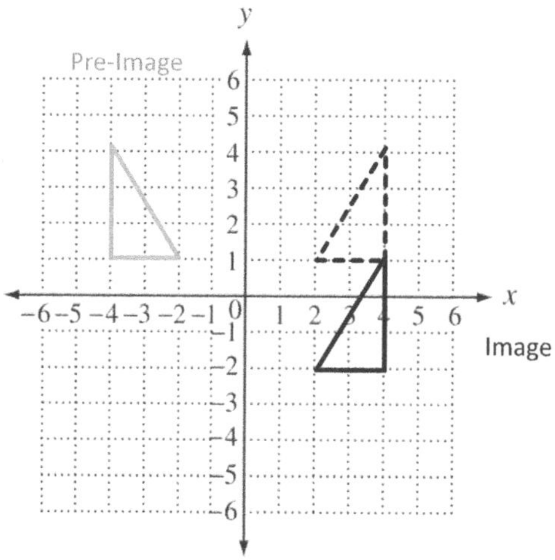

Dilation: increasing or decreasing the size of a pre-image without changing its shape or angles.

Example: Dilate the triangle by a scale factor of 2.

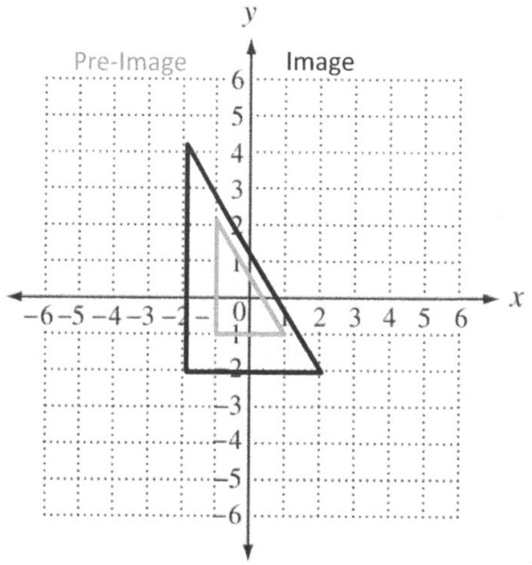

Made in the USA
Monee, IL
08 November 2026

68002503R00079